CAREER AS AN

AIRCRAFT MECHANIC

AVIONICS TECHNICIAN

YOUR EDUCATION, YOUR TALENTS AND skills, your hobbies and interests, and your desires for the future will all come together when you take that first step down the career path of your choice.

It may be true that pilots are the first thing you think of when considering careers in aviation but pilots would not stay in the air for very long without mechanics and avionics technicians. Aircraft mechanics are responsible for maintaining the aircraft engines and airframes, while avionics technicians take care of the electronic systems inside the plane. Aircraft are very maintenance-intensive. They are subject to much greater stresses than cars, for example. When cars break down, drivers can coast onto the side of the road and call a tow truck. When aircraft break down they can fall out of the sky.

Aircraft mechanics and avionics technicians are very highly skilled and highly valued. Pilots may get the recognition from the public, but pilots know the importance of the mechanics and technicians who service their planes. When an aircraft mechanic says there is a problem, everybody waits for the mechanic to fix it. If you have ever experienced a flight delay and sat in the airport fuming, you can thank an aircraft mechanic. Without these skilled professionals you would never get to your destination at all. Airlines, parcel delivery services, freight carriers, the military, and every other enterprise that uses aircraft all depend on mechanics.

WHAT YOU CAN DO NOW

THE SKILLS NEEDED TO MAINTAIN AIRCRAFT are very similar to those required to repair automobiles. Take shop classes while you are still in high school. Better yet, buy an old car and fix it up. Nothing will teach you how a complex mechanical system works better than taking it apart and putting it back together again. Total auto restoration projects can take months or years to complete, but small projects can keep you happily occupied for a weekend. Some time spent under the hood of a car will teach you how to use tools correctly and give you a preview of the work you may be doing someday.

If you are at least 17 years old you can earn a private pilot's license. Flight training can be expensive, but if you can afford the time and money it is an experience that will serve you well for the duration of your career. To become a private pilot you will have to complete at least 40 hours of training from an instructor certified by the Federal Aviation Administration, or FAA (you will soon find that the FAA establishes most of the rules for the aviation business). As a private pilot you cannot get paid to fly aircraft, nor can you fly after dark without earning additional certifications. You can, however, rent a plane for a few hours and take your friends on a short trip. It is not a requirement to learn to fly a plane in order to become an aircraft mechanic, but many mechanics do earn private pilot's licenses. Doing so gives them a better understanding of airplanes and how they do what they do.

There is no better, faster way to learn about an occupation than by reading the trade journals devoted to it. Many magazines cover the aviation business. Among the more popular titles are *Plane & Pilot, Flying, Flight Journal, Aviation Safety and General Aviation News.* Pick a few that you like and subscribe to them.

HISTORY OF THE CAREER

WILBUR AND ORVILLE WRIGHT, INVENTORS of the first successful airplane, were bicycle mechanics. Their first plane resembled a bicycle in many ways. Spindly, lightweight and held together with cables like spokes, the Wright Flyer was the first aircraft to achieve true flight. Previous aircraft had all been gliders that were hard to control and usually landed on surfaces below the level from which they took off – they tended to glide downhill, which is not really flying. The Wright Flyer was powered, could be steered in any direction and landed at a point as high or higher than the one it took off from. The Wright brothers achieved this historic success on December 17, 1903 on the beach at Kitty Hawk, North Carolina.

The Wright brothers built the Wright Flyer from ordinary bits and pieces of wood, cloth and metal, and a lightweight aluminum engine, at a total cost of less than $1,000. Their mechanical skills revolutionized the world. Within 15 years, aircraft evolved from the spindly Flyer into the dashing and aerobatic fighter planes of the First World War. This lightning-fast innovation was spurred almost entirely by mechanics always trying to make something better.

The Wright Company, founded in 1909 after the Wright Brothers had achieved some success with their earliest designs, lasted only until 1915. Aviation technology matured rapidly, which created demand for new types of engineers who could design complex aircraft. Mechanics, for the most part, got out of the design and inventing process, but were needed in ever greater numbers to maintain and service the very demanding new flying machines. Aircraft proliferated quickly after World War I, with government agencies like the Post Office Department, and the Army and Navy funding research and buying aircraft, and passenger airlines offering flights to early adventurous travelers.

The federal government started to regulate the aviation industry in 1926 with the passage of the Air Commerce Act. Among other things, the act charged the Department of Commerce with licensing pilots, certifying aircraft, operating and maintaining aids to airborne navigation, and generally making and enforcing the rules regarding aviation traffic and safety. The department also started requiring certification for mechanics maintaining aircraft. In 1938, these functions were transferred to a new agency called the Civil Aeronautics Authority, or CAA.

Aircraft design advanced very rapidly in the 1920s and 1930s. Propelled by the technological leap provided by the war, aircraft became faster, sturdier, and more reliable. Mechanics could no longer improvise. They needed professional skills and training to master the increasingly complex machines.

The next big leap in aviation technology resulted from World War II. All sides invested heavily in aviation technology during the war, devising some of the most famous aircraft ever built, including the British Spitfire, American P-51 Mustang, and Japanese Zero. The first long-range bombers also came into use, like the British Lancaster and the American B-17. The Navy and Army Air Corps, forerunner of the United States Air Force, trained and deployed mechanics wherever they sent aircraft – thousands of mechanics.

The war changed the world of aviation in several ways. The urgent need to stay one step ahead of the enemy spurred extreme technological innovation in a short time, which often happens during wars. Much of that technology was easily transferable to the civilian sector, including using long-range bomber technology to build the first large passenger and cargo planes. After the war thousands of trained mechanics and pilots flooded into the job market, creating a ready pool of talent for private companies eager to turn the bomber assembly lines into lines churning out passenger and cargo aircraft. The ensuing Cold War with the Soviet Union ensured that the military would continue to train mechanics and pilots for decades to come, providing an almost-free training pipeline for the civilian aviation industry. That training pipeline is still a critical component of the aviation industry today.

The CAA became the Federal Aviation Agency in 1958. The name was changed to the current Federal Aviation Administration in 1967, when the FAA was incorporated into the new Department of Transportation. The FAA certifies mechanics and avionics technicians under Part 147 of the Federal Aviation Regulation or FAR. Talk to mechanics and pilots and they will routinely refer to the "FARs" when the topic turns to regulatory matters.

Early avionics – a combination of the words "aviation" and "electronics" – were largely a product of Cold War competition with the Soviet Union. Small radar sets were installed in some aircraft as early as the late 1930s, but the technology did not become common in aircraft until the 1950s. Along with radar came sophisticated communications equipment, navigation systems, aircraft engine management and flight control systems, collision-avoidance and weather systems, and, for military aircraft, a host of targeting and surveillance tools. Today about 20 percent of the cost of a fighter plane is represented by avionics. Furthermore, most modern aircraft would have a very difficult time flying or landing safely without avionics equipment. The electronics have become inseparable from the airframe and engine.

There are approximately 125,000 aircraft mechanics in the United States, and about 18,000 avionics technicians. The largest employers of aircraft mechanics and avionics technicians are commercial airlines and cargo services, specialized service companies, the federal government, and aerospace manufacturers.

WHERE YOU WILL WORK

AIRCRAFT MECHANICS AND AVIONICS technicians can find work wherever there are aircraft. In the United States the FAA regulates most aspects of the aviation industry and requires that all maintenance more complicated than oil changes be carried out or supervised by mechanics and technicians with proper credentials.

There are several concentrated areas for aircraft mechanics and avionics technicians. Major metropolitan areas are home to many job openings because they tend to have large, high-volume airports and associated maintenance facilities. The airlines and service centers operating out of a major airport employ hundreds of mechanics and technicians. Smaller cities also have airports, of course, but there will be fewer opportunities there to get into the job market and to move up. There are exceptions to the rule. Wichita, Kansas, for example, is home to an impressive number of aircraft manufacturers and service centers. Wichita established itself as an aviation hub in the 1920s and grew rapidly during World War II, when the federal government encouraged manufacturers to expend operations in the middle of the country because it was easy to move finished aircraft to the coasts, and because it was assumed that the middle of the country would be safe from attack.

The United States military is home to many aviation jobs – tens of thousands of them. All of the services use aircraft. The Navy and Air Force operate the largest number of aircraft by far, but the Marine Corps and Coast Guard have small fleets of their own, and the Army operates thousands of helicopters. Many aircraft mechanics and avionics technicians start their careers in the military. With a five-year commitment, the military will send enlistees to school to learn their skills, and then sends them into the field to work on the most sophisticated aircraft in the world. Many careerists stay in the military for 20 years or more and earn a pension, but even those who get out after one hitch are snapped up by the aviation industry. They come fully trained, highly disciplined and ready to work. The military is not for everybody but you should definitely consider starting your career there.

Aviation is the fastest form of transportation. Airlines move their people around every day, and nobody thinks twice about living in one city and working in another hundreds of miles away. Mechanics do not have quite the same arrangement as flight attendants and pilots, who are in constant motion, but mechanics can be moved around, too. If you live in Chicago and an aircraft in Dayton needs your skills you may go to Dayton for the day.

DESCRIPTION OF WORK DUTIES

Civilian Aircraft Mechanic
Most aircraft mechanics are civilians who work for private companies like airlines and aerospace manufacturers, or for various government agencies. Some work for the military, but in a civilian capacity. They must all be certified by the FAA.

In order to start working, aircraft mechanics have to earn at least one rate certification, but most earn two. The basic rates are airframe and powerplant, called "A and P." Both require mechanics to pass written, oral and practical examinations. Before you can take the exam you must accrue at least 18 months of experience working with airframes or powerplants, or 30 months of working with both simultaneously. You must also be at least 18 years old and fluent in English. Graduating from an FAA-approved aviation maintenance technician school can substitute for the experience requirement, and usually takes about 18 to 24 months of combined work and study. Military training in certain military occupational specialties also confers eligibility to take the exams.

The exceptions to the rule possess what is known as a repairman's certificate, which is a sort of steppingstone to full certification. In addition to being at least 18 years old and fluent in English, careerists who want a repairman's certificate must be employed in a job that requires them to work on aircraft, have at least 18 months of practical experience or a diploma from an FAA-certified school, and be recommended by their employer. Repairmen, however, must perform their duties under the direction of an FAA-certified supervisor and cannot sign off on repairs. Most repairmen are on their way to full certification.

Once certified, A and Ps can work on airframes and powerplants on their own and can approve repairs and return aircraft to service.

Different aircraft or specialized systems often require their own certifications, leading many A and Ps to earn literally dozens of certificates over the course of a career.

Aircraft essentially consist of three major systems: airframes, powerplants, and avionics. Airframes include the plane's fuselage, wings and control surfaces. Think of this as bodywork. Unlike auto bodies, where much of the bodywork is basically stylish packaging that does not have to bear much stress, aircraft fuselages have to withstand enormous stresses. Wings provide the lift that allows the aircraft to fly, fuselages are pressurized, and control surfaces allow the pilot to control the plane. All of these elements are twisted and torqued every which way during a flight and they have to hang together to support the plane in the air, and while taking off and landing. One of the first commercial jetliners, the De Havilland Comet, suffered a series of crashes in the 1950s in which the planes literally broke apart in the sky due to operational stresses. The cause was found to be square windows that caused metal to fatigue at the corners. The Comet is why aircraft windows have had rounded corners ever since.

Powerplants are the engines that give aircraft the thrust to put to use the lift generated by the wings. Engines range from simple piston engines not much different from those used in cars, to enormous jet engines built by General Electric and Rolls-Royce. As with airframes, different engines may require special certifications.

Today's jet engines are incredibly reliable and efficient. Most people think aircraft made the move from propeller-driven piston engines to jet engines because jets are faster. Jets are faster, but their biggest appeal was that they are so much more reliable than piston engines. They have fewer moving parts and spend less time on the ground and more time in the air.

A and Ps with at least three years of certified experience may apply to the FAA for an inspection authorization, or IA. A and Ps with IA may inspect aircraft and recommend repairs and maintenance. They may also approve all work done and authorize aircraft to return to service.

Civilian Avionics Technician

Avionics are the third basic element of modern aircraft. A combination of the words "aviation" and "electronics," avionics includes all of an aircraft's electronic systems, which can be very extensive and complex. A modern passenger plane relies upon an extremely sophisticated suite of electronics, including radar, communications, flight control systems, guidance systems, and automatic pilots, among others.

In order to become certified as an avionics technician, you must first earn an FAA airframe certification. In practice, however, most careerists earn both A and P certifications. They must also earn a radiotelephone operator's license from the Federal Communications Commission. These certifications allow them to enter the profession, but they will not get very far without additional training and certification. Certifications are earned according to the needs of an individual job, but additional training can be pursued at any time. Many avionics technicians earn an associate or bachelor's degree in electronics or avionics.

Modern avionics are truly amazing. Much of today's working technology was the stuff of science fiction only a couple of decades ago. Civilian airliners essentially fly themselves, relying on pilots to monitor systems and handle takeoffs and landings. Jets are vectored into landing slots and runways by avionics systems networked with the control tower and other aircraft in the area to make sure that aircraft stay out of each other's way. While in transit aircraft are in constant contact with FAA ground control stations. Planes can find their way around the globe using signals from Global Positioning System satellites.

Military Aircraft Mechanic
Each of the military services trains and deploys its own aircraft mechanics. Military aircraft mechanics work on the most sophisticated aircraft in the world, often under very challenging conditions.

In order to join the military, potential enlistees must first take the Armed Forces Vocational Aptitude Battery, an exam commonly known as the ASVAB. An enlistee's ASVAB score determines which military occupational specialties – known as MOSs – he/she qualifies for. Each service has different requirements, but if you want to come anywhere near military aircraft you will need to do well on the test.

Most MOSs are agreed upon in the recruiting station. After signing an enlistment contract enlistees ship off to boot camp for nine weeks, and then to a school to train them to become aircraft mechanics. After completing training they can be deployed to squadrons anywhere in the world, on land or at sea. After a few years of work experience they will return to the classroom for additional training. They may also take on managerial responsibilities as they move up in rank. This leadership experience is one of the reasons military veterans are readily employed in the private sector when they return to civilian life.

A complete list of the aviation MOSs offered by the services would fill the rest of this report. A small sampling is presented here. For more information go to each service's recruiting website. Addresses can be found at the end of this report.

Air Force mechanic MOSs include aerospace maintenance, aerospace propulsion, and aerospace propulsion with a specialty in turboprop and turboshaft powerplants.

The Army offers MOSs in aircraft structural repair, and specialized opportunities for each type of helicopter and unmanned aerial vehicle in the Army inventory.

The Coast Guard trains aviation maintenance technicians to maintain its fleet of fixed- and rotary-wing aircraft.

The Marine Corps certifies helicopter and fixed-wing airframe and powerplant mechanics, and specialists in tilt-rotor technology.

The Navy, which calls its MOSs "rates," trains sailors to become aviation machinist's mates and structural mechanics on everything from C-130 cargo planes to fighter planes and helicopters.

Military Avionics Technician   Military avionics uses the most advanced technology in the world. Much of it is classified and top-secret. If you want to learn more about it you will have to enlist. A good reason to stay out of trouble. The services will not grant a security clearance to enlistees with police or financial problems.

Military avionics can guide fighter aircraft to a pinpoint target thousands of miles from where they took off. Avionics can jam enemy radar and even send confusing signals to make a large aircraft look like a small aircraft, or even a flock of birds. Military avionics uses lasers to guide weapons to targets in the air and on the ground.

Air Force avionics MOSs include avionics systems and airborne intelligence, surveillance, and reconnaissance. The Army trains general avionics mechanics and specialists for each type of helicopter. The Coast Guard trains avionics electrical technicians. Marine Corps avionics MOSs include aircraft electronic countermeasures technician and aircraft systems technician for each type of aircraft. Navy avionics rates include aviation electronics technicians and aviation electrician's mates.

I Am an Aircraft Service Technician

for an Airline "I always knew what I wanted to be when I grew up. I wanted to be one of the lucky few who get to work on airplanes. I was fascinated by them as a kid, and I've never grown out of it. Airplanes are exciting, and they always will be.

I did a hitch in the Air Force as a mechanic working on C-17 cargo aircraft. Very large, very long-range aircraft that are deployed all over the world. Not quite as numerous as their smaller counterpart, the C-130, C-17s are notable for the fact that they can carry several times as much cargo as a C-130 but don't need much more room to land. That's a nice thing when your job is to move people and cargo around the world at a moment's notice.

I did a five-year hitch and immediately turned my experience into A and P certification. A major airline hired me as soon as I passed the exam, and I started my civilian career. I now have enough specialized certificates to wallpaper a room. I'm very proud of each and every one of them.

I work on large, long-haul aircraft – Boeing and Airbus planes that seat hundreds of passengers and regularly fly across the Atlantic and Pacific oceans. These planes are behemoths. It is amazing that they get off the ground.

The airline business is driven by efficiency. Airplanes don't make money when they're on the ground. The goal is to keep them in the air as much as possible. Over the years the planes have become extremely reliable.

I work mostly on engines. Our fleet has engines built by General Electric and Rolls-Royce, the two largest makers of jet engines. I can actually walk around inside the latest engines. In the past, long-haul aircraft were required to have four engines to make sure that they would always have a spare in case something went wrong. Today's engines are so reliable that even the biggest planes often get only two engines. They are so big you have to see them to believe it. The plane's computers can program the control surfaces to keep flying in a straight line even if one engine fails.

New aircraft may be incredibly reliable but they only stay that way with constant attention from mechanics. We have our own maintenance schedules that go beyond FAA and manufacturer requirements. We get to know our aircraft very well, and take the time to seek out trouble spots. Our reputation is on the line.

I'm gradually cashing in my GI Bill benefits to earn a bachelor's degree in aviation maintenance technology. Eventually I plan to move into management, but I'm in no hurry. I take great pride in maintaining a safe and reliable fleet."

I Am an Inspection-Authorized
Mechanic for a Shipping Company

"My employer owns one of the largest fleets of aircraft in the world. We are bigger than most passenger airlines and account for more landing slots at our hub airport than any other operator. We deliver millions of packages around the world every day.

I was an A and P for about 10 years before I requested an inspection authorization. I liked being a mechanic, but as I got more senior I had an opportunity to move into a management position, which required the authority to inspect aircraft. I never thought I would make the transition into an inspector. When I graduated from A and P school all those years ago I thought I would be a mechanic forever. I love to tinker, and earning additional certifications for new planes was one of my favorite things.

My main responsibility is to conduct inspections according to schedules established by the FAA and the manufacturers. Aircraft inspections are very carefully documented, and documents are routinely inspected by the FAA. Nothing is left to chance. FAA regulations say that all aircraft, no matter how small, must be inspected annually. All aircraft that carry paying passengers or otherwise engage in commercial activity have to be inspected at least every 100 hours of flying time. If you hang around pilots and mechanics you'll hear us talk about hours a lot. Hours of flying time are the standard measurement in this business because aircraft only accumulate wear and tear when they're in the air. The passage of time doesn't matter anywhere near as much as hours in the air.

I am one of a handful of inspectors responsible for the clutch of aircraft based at our facility. We inspect aircraft on schedule and work closely with the mechanics to make sure that all work is performed properly. The FAA and the manufacturers both set standards and they must be met every time. When work is completed we can authorize the aircraft to return to service. It's a very large responsibility, and one not taken lightly.

I would recommend this career to anybody who is mechanically inclined, detail-oriented and really crazy about airplanes. The aviation business has always been characterized by people who are passionate about what they do. If you're one of them you'll fit right in."

I Am an Aviation Electronics Technician (AT) in the Navy

"I am responsible for maintaining the avionics systems on some of the most sophisticated aircraft in the world. I can't even talk about some of the things I do because they are classified. Let's just say they are very, very cool.

I joined the Navy right out of high school. I did well on the ASVAB, so I pretty much got to choose my rate. Avionics sounded interesting, so I signed the dotted line and headed off to AT school after boot camp. AT is the abbreviation for my rate. I am a petty officer, first class so my title is AT1. The next step is to make chief petty officer, which is quite an achievement.

I've done my job all over the world. I did a few tours in combat zones in the Middle East. 'The sandbox,' as we call it. This may come as a surprise, but I actually enjoyed it. Sure, living in a tent gets old, and the constant uncertainty is a bit stressful, but it's also invigorating. There is nothing like being at the tip of the spear to remind you why you joined.

I've also served aboard aircraft carriers. I've worked on FA-18 fighter jets. There is nothing quite like a carrier deployment. It was just me and 5,600 of my closest friends. In that job I got to work on radar, navigation, communications, fire control systems, tactical displays, electronic warfare systems and even antisubmarine warfare sensors. My colleagues and I did all of our work in the hangar bay, the cavernous space below the flight deck where the planes are parked for servicing.

Through the GI Bill I've been able to earn an associate degree in avionics systems. I've even taken classes while deployed. I'm working on a bachelor's degree now, and will finish in a couple of years. I haven't decided how long I'll stay in the service. I thought I would just do five years and get out, but I'm at the 10-year mark now and I like it much more than I ever thought I would. Some of my friends have gotten out, and most of them had jobs within a matter of weeks. Airlines actively recruit military veterans with the right rates or MOSs."

I Am a CH-47 Helicopter Repairer in

the Army "The Army owns a very small number of fixed-wing aircraft but thousands of helicopters. We use them for everything, from moving soldiers into combat zones, to general transport, to medical evacuations, to cargo. We also use attack helicopters to bring the fight to the enemy. Attack helicopters are like airborne tanks.

My specialty is the CH-47 helicopter, usually known as the Chinook. The Chinook is a twin-rotor helicopter well-suited to a variety of missions. Mostly, it is used to move large numbers of soldiers or supplies around combat zones. Its range is excellent and it can carry a large load, for a helicopter. The CH-47 has been in service since the early 1960s and has proven to be a very reliable airframe. Most aircraft are retired from service after a couple of decades, but the Chinook keeps soldiering along. Powerplants and avionics have been upgraded several times over the years, and the Chinook remains a powerful tool for the Army.

In a typical day I may inspect the fuselage and make repairs, service and lubricate various systems and subsystems, and maybe even pull an engine for a rebuild. I spent 16 weeks in school after basic training to learn the skills necessary to work on the CH-47. Those 16 weeks were just an introduction. I've learned most of what I know by working alongside veteran repairers who have seen it all. No classroom can prepare you for the mess of an engine choked with dust from the Afghan desert. That's something you have to learn on the job.

I know a great job will be waiting for me when I decide to get out of the Army and return to civilian life. Fixed-wing aircraft are more common than helicopters, but helicopters are much more maintenance intensive. Nothing needs more maintenance than a helo. That's what I call job security."

I Am the Owner of an FBO Aircraft
Maintenance Company

"Only people in the aviation business know what an FBO is. FBO stands for fixed-base operator, and refers to a business that has been granted permission to set up shop on the grounds of an FAA-certified airport. Essentially, FBOs are repair shops and fuel stations that serve aircraft based at the airport or which are passing through. Even very small civilian airports have an FBO, if only to sell fuel to passing pilots. Uncontrolled airstrips may not have FBOs, but they usually don't have permanent hangars, either.

I got into this business after a couple of decades working for somebody else. I earned my A and P certifications after high school and started working for a small regional airline. I learned a lot about small jets, the kind that fly passengers relatively short distances on domestic routes. Even the smallest jets are technological marvels. To this day, when I look at a jet I see an airborne hot rod.

I moved up to a major airline for a few years and learned about long-haul jets. The big ones that move hundreds of passengers long distances. I decided I liked smaller planes and the kinds of people and business that they attract, so I got into corporate aviation. Many small airports are home to jets used by businesses to fly their sales staff and senior executives to the places they need to be. Corporate aviation is a huge business that outsiders hardly know about. I worked for an FBO at a small civil airport. We worked on planes owned by several very large corporations and a few small, local companies. We also got a steady stream of business from private pilots who owned aircraft based at the airport. I found it to be a very chummy bunch of people who genuinely enjoy flying.

I started my current company with a couple of partners a few years ago. We serve a small civilian airport on the fringe of a major metropolitan area. The hangars around our maintenance facility house aircraft owned by some very familiar names. Those aircraft are critical to the success of their companies. Some people like to dismiss corporate aviation as being for fat cats who get to fly around in style. Nothing could be further from the truth. When a corporate executive hops in a small plane, flies to a meeting, makes a deal that keeps hundreds or thousands of people employed, and returns to the home office by mid-afternoon, everybody wins.

I never earned a degree in business administration. I learned everything I know by paying attention and never being satisfied with anything less than my best effort. This is a very interesting, complex business. It attracts interesting, complex people, too. That's my favorite part."

PERSONAL QUALIFICATIONS

ANYBODY CAN LEARN THE TECHNICAL skills necessary to become an aircraft mechanic or avionics technician, but most people would not succeed in these jobs without a few personal qualifications.

Analytical skills are very important, and so is an eye for detail. All mechanics and technicians are problem solvers. It does not matter if their specialty is aircraft, automobiles, or toaster ovens. Their job is to sort things out and get to the bottom of problems. Mechanics and technicians spend most of their time conducting routine maintenance and testing – this is fairly straightforward. They prove their worth when things go wrong. This is when keen analytical skills come into play. Mechanics and technicians must be able to assess a problem one step at a time, analyzing everything that could have gone wrong until they find the right solution. This process can require great ingenuity when the problem is not obvious. Try tracing a short in an old car and see how far you get before you get frustrated.

Having a good eye for detail is related to analytical skills but it is not quite the same. Often it is a good eye for detail that reveals the problem that needs to be solved. Aircraft systems rarely fail outright. In addition to being very sturdily built, most aircraft systems are redundant. That means they have backup systems built-in. When one fails the other kicks in and the plane keeps flying. Computers tend to monitor most of the critical systems but they cannot do everything. The mechanics and technicians conduct routine inspections of all airframe and avionics systems in order to forestall any breakdowns. This can be as simple as noticing that a bolt is loose or that something is out of place. Many professions require people who are detail-oriented, but this is seldom as critical as in the aviation industry.

Physical dexterity is a must. Mechanics will tell you all about their latest aches and pains, bumps, scratches, and bruises. There will be stories of how they wormed their way into spaces not intended for humans. Aircraft mechanics can find themselves literally crawling into engines and fuselages to get to the things they need to fix.

ATTRACTIVE FEATURES

THE AVIATION INDUSTRY IS ONE OF THE largest and most important economic forces in the world. Aviation is definitely an exciting industry to be a part of. Aviation has its fans and its legends, and is influential in all our lives. The cars of the 1950s and 1960s, for example, were designed to look like aircraft, with taillights that resembled rocket engines and enormous tail fins standing in for wings. That was because flight has always played an outsized role in the human imagination.

In a more practical sense, aviation is also one of the key industries that stimulates the global economy. Only about 10 percent of commercial goods travel by air – the rest travel by sea – but hundreds of thousands of people travel by air every day. This is especially critical in an economy based on services, where people are the most important assets. Being a part of such an important industry is something to take pride in.

Mechanics and technicians tend to love what they do. Ask mechanics about their work and how they feel about it. You will find that mechanics who specialize in Chevrolets, for example, have a very real love and enthusiasm for Chevrolets. There is something about being a technical specialist that generates enthusiasm like nothing else. Mechanics who work on Boeing, Piper or King Air aircraft feel the same way. They become experts in their fields and take great pride in knowing everything there is to know about their specialty.

People who are cut out to be technicians and mechanics almost always do it because they enjoy it. They like tinkering and are satisfied when they do a good job. If you are one of those people you will never be happier than when you are elbow-deep in a project.

This field presents interesting opportunities. You can spend a few years in the military learning your skills. You do not have to stay for 20 years, but you can certainly stay in long enough to get some first-rate training and see some of the world. Outside the military, aircraft mechanics and avionics technicians can work in some very interesting places. Remember, wherever there are planes there are mechanics and technicians. Want to work in Africa or the Middle East for a few years? How about in a testing facility experimenting with new technologies? Aviation is a cutting-edge industry. In fact, being at the pinnacle of technology has been one of the industry's hallmarks right from the beginning.

UNATTRACTIVE ASPECTS

THE AVIATION BUSINESS IS SUBJECT TO serious ups and downs in the larger economy. The fact that aviation is critical to the overall economy is both a blessing and a curse. It is a positive that the business will always move forward, will always advance technologically, and will always grow along with the rest of the economy. It can be a problem when the larger economy turns down, because so does the aviation business. When things get tough, people and businesses cut costs where they can. Travel is often the first expense to be reduced. Corporate managers may decide to hold a few meetings via teleconference and reserve travel only for the most critical engagements. Families may decide to take driving vacations instead of flying long distances in order to save money.

Airlines are not the only employers for aircraft mechanics and avionics technicians. The economy does not affect the military much, and more essential operations like cargo delivery are not affected to the same degree as passenger airlines. Still, it is a little-known fact that there are hundreds of passenger jets parked in the desert in several southwestern states (where they will not rust) that are leased by the major airlines for short periods of time to accommodate fluctuations in passenger traffic. When demand goes down, airlines shed aircraft, and fewer aircraft require fewer mechanics to maintain them.

Today, an American can buy a round-trip ticket to Europe for less than $1,000, cross the Atlantic in a few hours and not be bothered by anything more annoying than jet lag. When Christopher Columbus crossed the Atlantic in 1492, he needed three ships with crews and the voyage was so expensive that it required financing from the king and queen of Spain. Few people thought he would make it. The fact that hundreds of thousands of people crisscross the world every day without a second thought is remarkable. This is made possible by jet-powered aircraft that were science fiction only a few decades ago. Because safety and expense are the paramount concerns in air travel, manufacturers have made great strides in improving efficiency and reliability. This all comes at a cost for mechanics and technicians. Even though air travel is expected to grow, demand for aircraft mechanics and technicians is expected to remain relatively flat. Modern aircraft need less and less hands-on maintenance.

People who work with their hands have always been looked down upon by some. People doing mechanical work are perceived as less important. If all the world's Shakespeare scholars disappeared tomorrow almost nobody would notice. If the world's aircraft mechanics didn't show up at work, the world would grind to a halt. Opinions to the contrary are nothing more than misguided snobbery.

EDUCATION AND TRAINING

COLLEGE IS OPTIONAL FOR THIS CAREER but studying is not. Aircraft mechanics and avionics technicians who want to move up in the world must be certified by the FAA. Those who want to branch out into specialties must earn additional certifications.

Technically speaking, aircraft mechanics and service technicians do not have to be FAA-certified to do their jobs. The FAA does require, however, that all work performed on aircraft be supervised by somebody with FAA certification. In practice, this means that certified mechanics have to look over the shoulders of non-certified mechanics to make sure they do everything right. This is acceptable for training a new mechanic but not for the long haul be-

17

cause it essentially requires two mechanics for every job. If you want to enter this profession, you will want to get certified.

There are two specialties within the field of aircraft mechanics: airframes and powerplants. These specialties are known as rates. To earn either rate a careerist must have at least 18 months of practical experience with either airframes or powerplants, or 30 months on both at the same time. The alternative is to graduate from an FAA-approved aviation maintenance technician program, which generally requires 18 to 24 months of study and hands-on training. Most careerists earn both airframe and powerplant certifications at the same time, commonly known as "A and P." Graduating from an approved program or demonstrating the necessary experience will qualify you to take the written, oral and practical exams that lead to certification.

Avionics technicians need to earn the airframe certification and a radio-telephone license from the Federal Communications Commission to get started on their career.

Many aircraft mechanics and avionics technicians start their careers in the military. All of the services operate aircraft. The Navy and Air Force operate the most by far, with the Air Force owning aircraft of all types, including enormous bombers and cargo aircraft, while the Navy's fleet – which is larger than the Air Force's – is long on small fighter planes and helicopters. The Marine Corps operates a small, highly specialized fleet of aircraft designed to support its assault mission, including Harrier fighters and Osprey transports that can take off and land vertically. The Army flies a huge number of helicopters and the Coast Guard operates helicopters and a variety of planes well suited to reconnaissance.

All of the military services train their own aircraft mechanics and technicians. Typically, new recruits go to nine weeks of boot camp followed by a longer course of training in aircraft mechanics or avionics. After completing initial training they are assigned to a squadron where they are put to work under the supervision of senior personnel. Assignments could be to anywhere in the world and can last from six months to three years. Additional training is required over time, all of which is provided by the service. The paychecks keep coming during training.

Military aircraft mechanics and avionics technicians get hands-on experience that more than satisfies FAA on-the-job training requirements to take certification exams after leaving the military. If you want to build a career in aviation you should definitely look into starting out in the military. One minimum five-year hitch can set you up for the rest of your career.

Some aircraft mechanics and avionics technicians do earn traditional college degrees. Many avionics technicians, in particular, earn associate degrees in electronics, or take specialized degree programs in avionics technology.

Bachelor's degrees are also offered in aviation systems management and aircraft mechanics for careerists who want to earn the extra credential. Some careerists also pursue degrees in aeronautical engineering, while others may add a degree in business administration so they can move into corporate management or start their own businesses.

All training programs and degree programs will come with hands-on experience. Students will work alongside established mechanics and technicians so they can put their classroom learning to use on real aircraft. In addition to being a great way to learn a trade, working with professionals in your chosen career is a great way to make connections that will serve you well after you have passed your exams and start looking for your first job.

EARNINGS

AVIONICS TECHNICIANS TYPICALLY EARN a little more than aircraft mechanics, but both professions pay well and offer plenty of room for advancement.

Aircraft mechanics just starting out can expect to be paid about $40,000 per year. Over time this figure can rise to about $75,000 and maybe more if you move into the managerial ranks or acquire additional certifications. Avionics technicians can expect to start their careers with slightly larger paychecks, about $45,000 per year. Over time, avionics technicians can earn about $75,000 to $85,000 per year if they earn additional certifications or take on managerial responsibilities.

Earnings are potentially limitless for entrepreneurs who start their own business, but so are the risks. When you are your own boss your success or failure depends solely on you.

Military personnel are paid according to a standard pay scale that applies to all the services. Pay is based on rank and time in service. Pay is supplemented by a housing allowance that varies depending upon where the service member is stationed. Military personnel also receive special pay for service in a combat zone and for having certain special skills. Junior personnel just out of boot camp and in the paygrade of E-2 are paid about $1700 per month. Junior personnel also receive free housing and meals. Within a few years, enlisted personnel can advance to the paygrade of E-6 and earn almost $3,000 per month after six years of service, not including a housing allowance and other allowances for which the service member may be eligible. For the complete military pay chart go to the Defense Finance and Accounting Service website at www.dfas.mil

Many aircraft mechanics and avionics technicians are represented by unions, particularly those employed by airlines. Unions typically push for higher wages and benefits, but they also collect dues from their members. Only you can decide if they will represent your needs.

OPPORTUNITIES

THE JOB OUTLOOK FOR AIRCRAFT mechanics and avionics technicians is generally good, although advances in the efficiency and reliability of planes mean that the field will have only average growth in employment, even though air travel is expected to continue to expand at a rapid rate for the next several decades.

Better planes use more complex equipment. That means that the mechanics and technicians who will get ahead are those who stay abreast of the latest technologies. Earning additional certifications is always a good way to move up in the world, and this will be especially true as aircraft become even more sophisticated. Credentials in related fields like computer science and engineering will also be helpful. Earning an associate degree or even a bachelor's degree will definitely help your career move along, especially if you have your sights set on management or on opening your own business.

If you walk into a recruiting office with an FAA certification already in-hand, you will go to the head of the line for the exact job you want. A good way to earn that certification is while serving in the military. No other employer will pay you to get training and then send you around the world to work on the most sophisticated aircraft in the most demanding conditions.

There are many other interesting employers who need capable aircraft mechanics and avionics technicians. Many nongovernmental organizations like Doctors Without Borders, the United Nations, and the Red Cross own and operate aircraft so they can reach remote corners of the world in an emergency. The military regularly hires civilian contractors to supplement uniformed personnel in missions around the world. There are many small airlines that fly to interesting destinations. Aerospace companies design and test the latest technologies every day, and they need mechanics and technicians to make it possible. This career rewards outside-the-box thinking and offers plenty of unconventional opportunities to get ahead.

20

GETTING STARTED

WHEN THE TIME COMES TO FIND YOUR first real job you should be ready to go. Get your personal marketing materials together. You should always have an up-to-date résumé on hand and ready to send out at a moment's notice. You never know what kind of opportunities may come your way, so it behooves you to be ready for them. You should prepare a properly formatted résumé even if you think there is little chance that you will ever submit it in paper form. Even though most job applications are filled out online, you may be asked to email a résumé to a potential employer who may then print it out. A properly prepared résumé will also make filling out digital forms easier because you can simply cut and paste relevant sections from your résumé into the form. Doing so will also keep you consistent and help you resist the urge to embellish or spin your résumé for different employers. Write a good résumé and update it whenever you do something new. If you do not feel confident in your own résumé writing skills you can hire a professional or turn to one of the many books or software applications on the market designed to help you.

Next, you can start contacting everybody you have ever met in the aviation business. These could be former instructors from an FAA-certified course, or aviation professors from college, or supervisors at the companies you worked for while you were completing your training. You might be surprised at how helpful these folks can be, especially if you made a good impression. It is very common for people to get their first jobs with the companies where they completed internships or on-the-job training. You are a known quantity, understand the company and its culture and, most of all, relieve the company of the need to spend time and money finding and training an untested applicant. Even if your connections do not have a job waiting for you they may know somebody who does.

In job hunting, you may get your hopes up one day only to have them dashed the next. Do not let it get you down. Opportunities are everywhere, and you will find one that suits you. At this stage in your career it makes no sense to hold out for the perfect job. You do not even know what the perfect job for you will be. Right now you will do yourself no favors by turning down opportunities where you could learn something and get experience. Seizing opportunities is what will eventually lead you to the perfect job and allow you to build the perfect career.

ASSOCIATIONS
PERIODICALS
WEBSITES

- aerotek
  www.aerotek.com

- Aircraft Mechanics Fraternal
  Association
  www.amfanational.org

- Aircraft Owners and Pilots
  Association
  www.aopa.org

- Alaska Airlines
  www.alaskaair.com

- American Airlines
  www.american.com

- AMT Society
  www.amtsociety.org

- Aviation Employment
  www.aviationemployment.com

- Aviation Institute of
  Maintenance
  www.aviationmaintenance.edu

- Aviation News
  www.aviationnews.net

- Aviation Pros
  www.aviationpros.com

- Aviation Safety Magazine
  www.aviationsafetymagazine
  .com

- Aviation Schools Online
  www.aviationschoolsonline
  .com

- AV Scholars
  www.avscholars.com

- Boeing
  www.boeing.com

- Delta Airlines
  www.delta.com

- Experimental Aircraft Association
  www.eaa.org

- Federal Aviation Administration
  www.faa.gov

- Flight Journal
  www.flightjournal.com

- Flying Magazine
  www.flyingmag.com

- General Aviation News
  www.generalaviationnews.com

- International Association of
  Machinists and Aerospace Workers
  www.goiam.org

- Jet Blue Airways
  www.jetblue.com

- Kitplanes
  www.kitplanes.com

- Liberty University
  www.liberty.edu

- National Business Aviation
  Association
  www.nbaa.org

- Plane & Pilot Magazine
  www.planeandpilotmag.com

- Plane Techs
  www.planetechs.com